math talk

mathematical ideas in poems for two voices.

by theoni pappas

Wide World Publishing/Tetra

4th Printing April 1999.

Wide World Publishing/Tetra
P.O. Box 476
San Carlos, CA 94070

Printed in the United States of America.

Library of Congress Cataloging-in-Publication Data
Pappas, Theoni.
Math talk : mathematical ideas in poems for two voices/by Theoni Pappas
p. cm.
Summary : Presents mathematical ideas through poetic dialogues intended to be read by two people.
ISBN: 0-933174-74-8 : $8.95
1. Mathematics--Poetry. [1. Mathematics--Poetry. 2. American poetry.] I. Title.
PS3566. A6229M38 1990
811' . 54--dc20 90-25380
CIP
AC

For Elvira & Kate

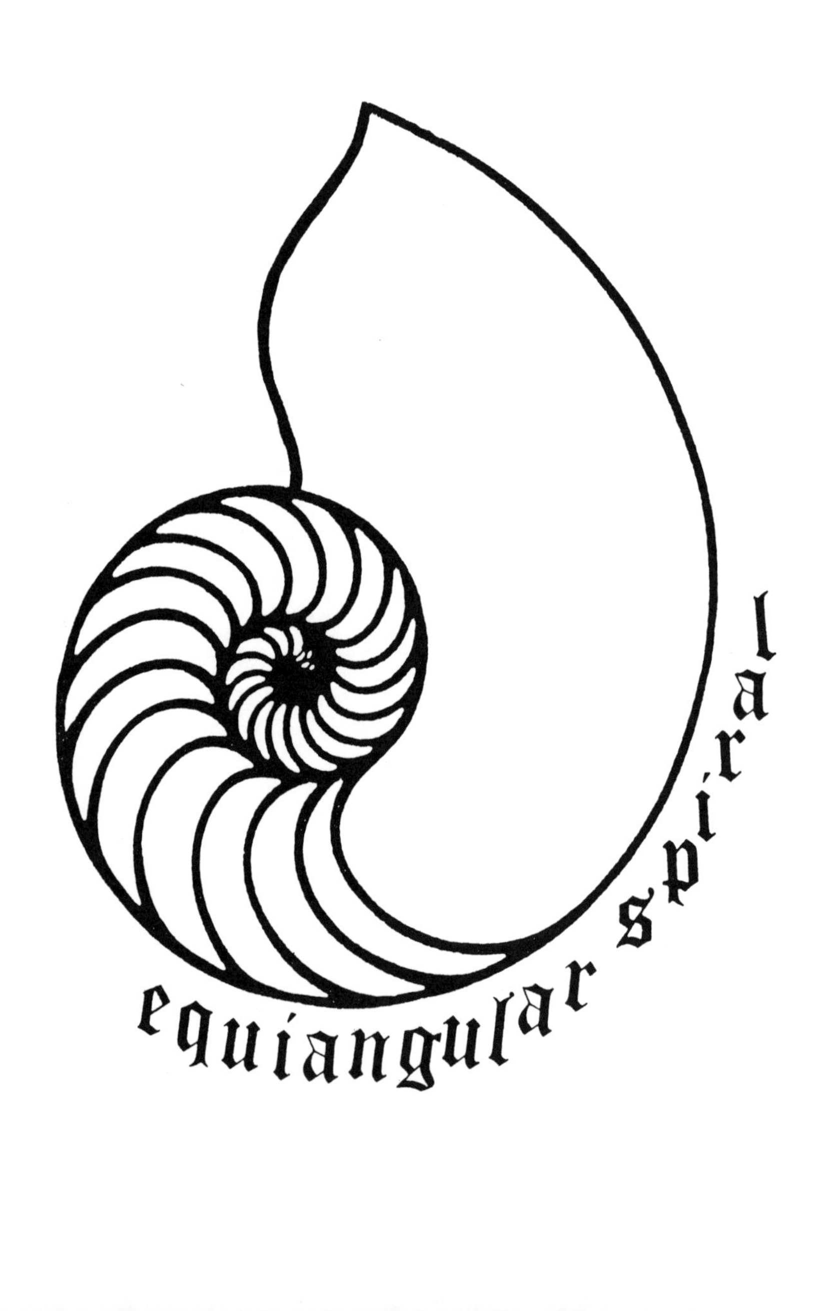
equiangular spiral

Contents

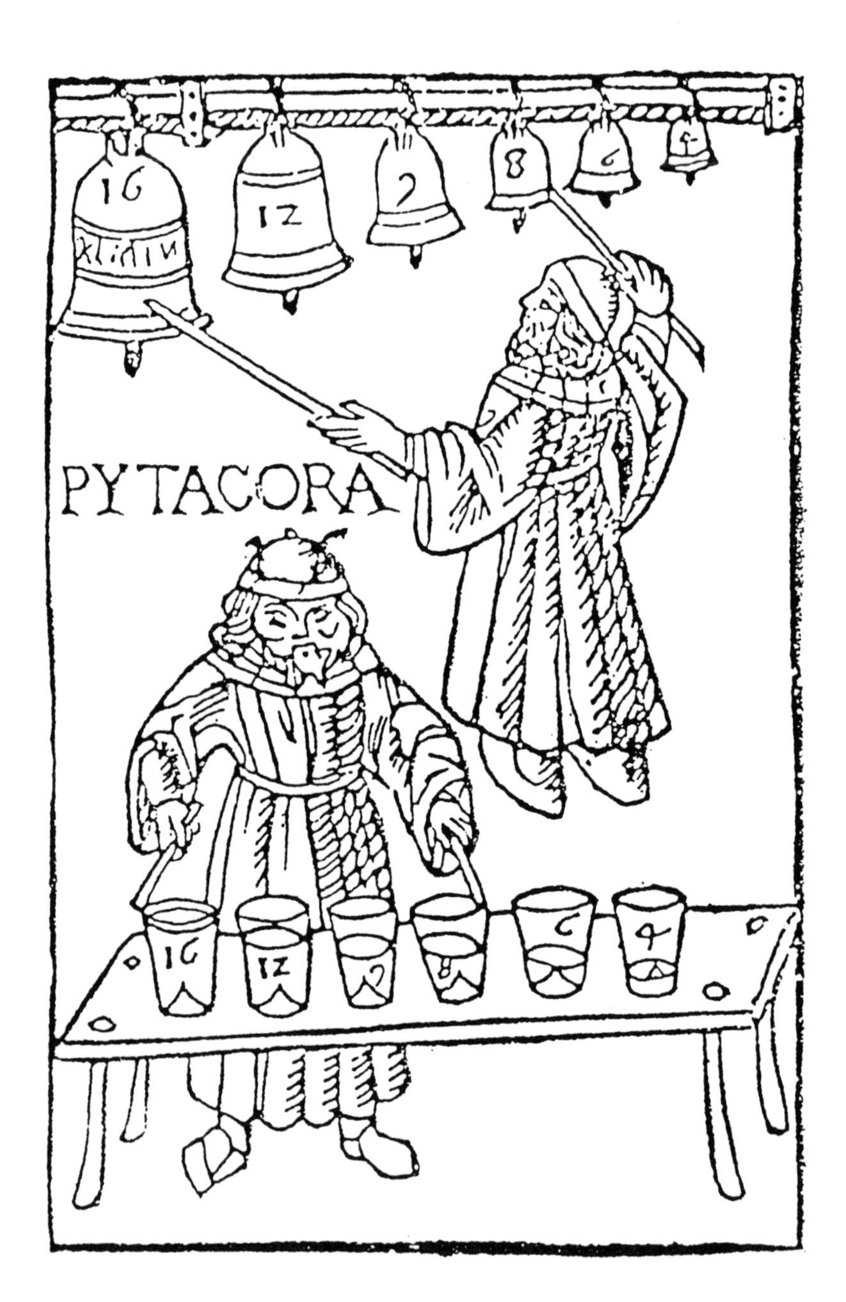

PYTACORA
16
xliiii
12
9
8
6
4
16
12
9
8
6
4

Forward

Mathematics may not seem to inspire poetry, since it is usually linked with the very logical.

Learning takes place via all our senses and by all forms of communication. Mathematical ideas can be learned through art, reading, conversations, lectures. Therefore, why not link mathematical ideas and poetic dialogues?

These poems in voices are meant to be read aloud. Each reader plays a role, thereby enhancing the entire piece. Please note that lines on the same horizontal are meant to be read simultaneously.

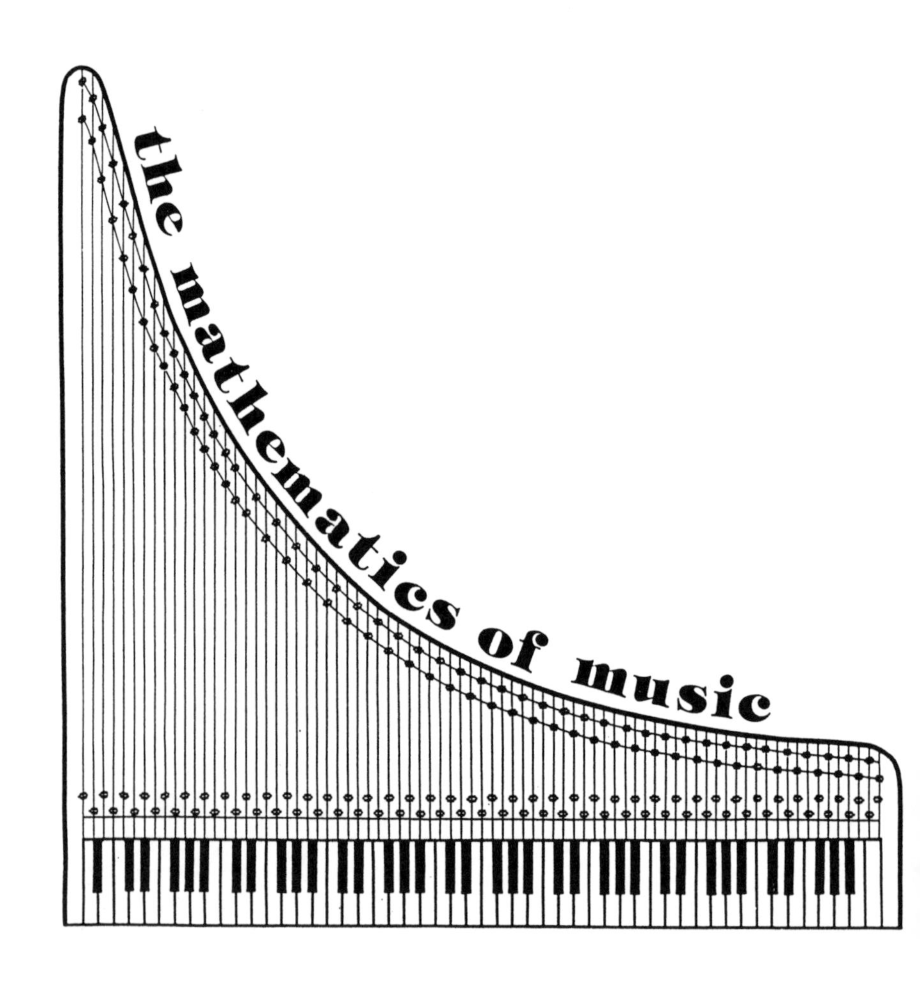
the mathematics of music

Mathematics

Mathematics

the word has been known

to conjure up

love — hate

delight — despair

recreation — anxiety

joy — fear

Mathematics

is a variety of subjects

and ideas

restricted only

by our imagination. by our imagination.

<table>
<tr><td>Computation,</td><td>Arithmetic,</td></tr>
<tr><td>geometry,</td><td>algebra,</td></tr>
<tr><td>trigonometry,</td><td>calculus,</td></tr>
<tr><td>topology,</td><td>tessellation,</td></tr>
<tr><td>hypercubes,</td><td>Klein bottles,</td></tr>
<tr><td>chaos theory,</td><td>Möbius strip,</td></tr>
<tr><td>fractals,</td><td>knots,</td></tr>
<tr><td>Pythagorean</td><td></td></tr>
<tr><td></td><td>theorem.</td></tr>
<tr><td></td><td></td></tr>
<tr><td>Mathematics touches</td><td></td></tr>
<tr><td></td><td>so much of our lives</td></tr>
<tr><td>of our world—</td><td>of our universe—</td></tr>
<tr><td>art,</td><td>nature,</td></tr>
<tr><td>music,</td><td>science,</td></tr>
<tr><td>history,</td><td>architecture,</td></tr>
<tr><td>economics,</td><td>literature.</td></tr>
<tr><td></td><td></td></tr>
<tr><td>Mathematics is</td><td></td></tr>
<tr><td>everywhere,</td><td>a pastime,</td></tr>
<tr><td>a recreation,</td><td>conundrums,</td></tr>
<tr><td>games,</td><td>puzzles,</td></tr>
<tr><td>problems,</td><td>solutions,</td></tr>
<tr><td>a way of thinking.</td><td>a way of thinking.</td></tr>
</table>

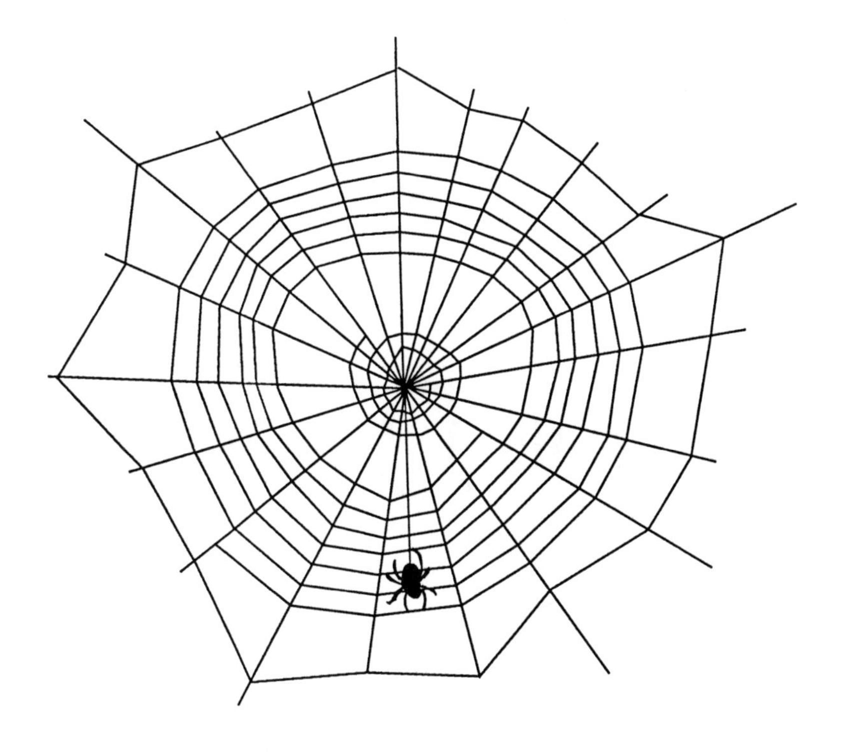

1

2

6

500

2,378

1,000,000

1/3

We are numbers

We are numbers.

Numbers.

First came one,

then two,

next three

and, four not far behind.

Five,

six,

counting naturals

large,

small,

five-hundred,

one-third,

gigantic,

minute,

a million,

one-millionth.

We count.
We count.

We add and subtract.

We multiply and divide.

We keep track.

We measure.

We are numbers.
We are numbers.

Large,

small,

never ending.
never ending.

0

-11

.03

π

9

√5

Circles

We like to think

of ourselves as perfectly round.

Not egg shaped

nor elliptical.

A circle's

never ending, nor beginning.

Always remaining

equidistant from its center.

Chords, tangents
radius diameter
π circumference
secants concentric
circumscribed inscribed

are some

objects and words that

adorn us.

Circles

never ending

yet finite in length.

The model

of the wheel.

The circle The shape

giving the most area

for a particular length.

Circles

Elegant and simple.

Circles Circles

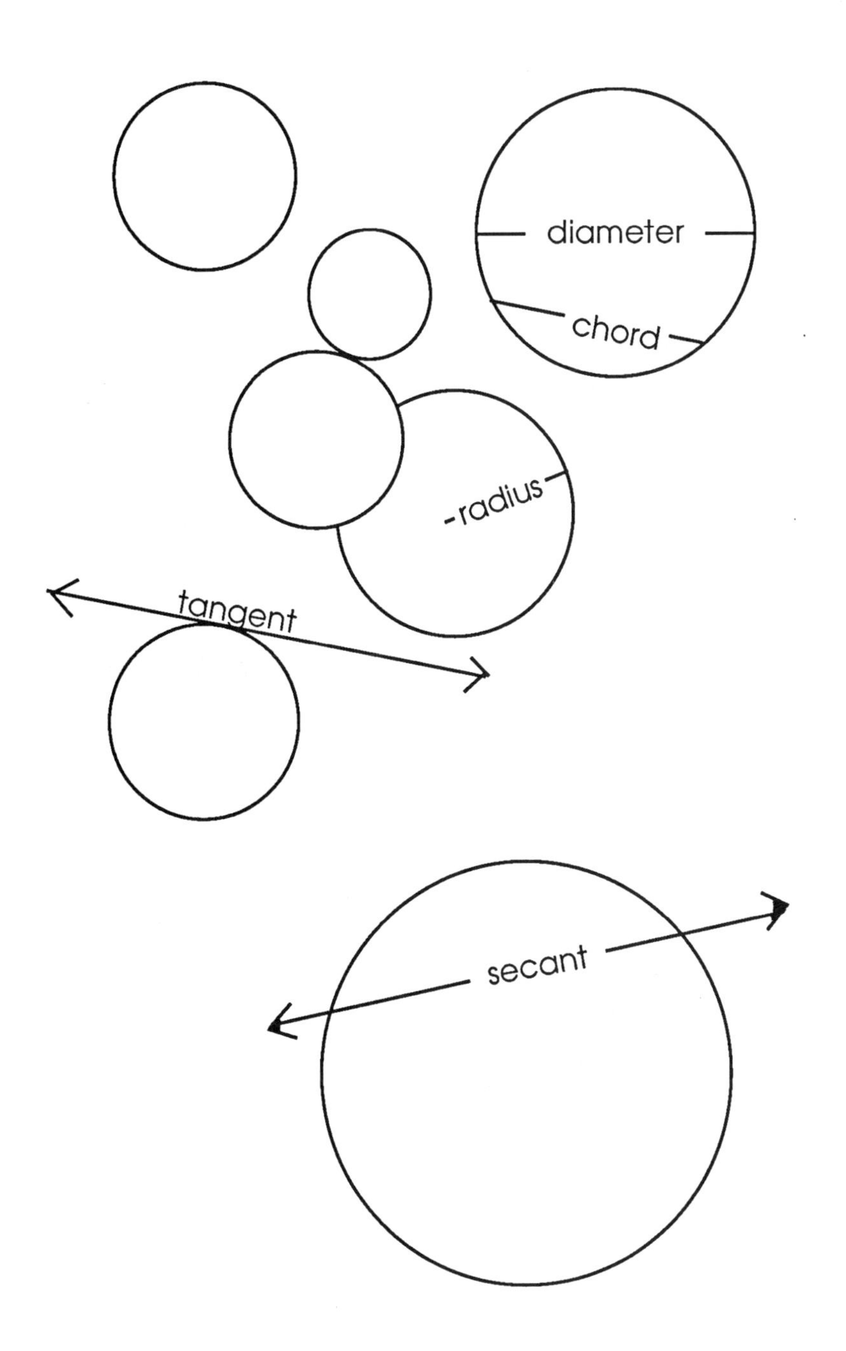
diameter
chord
radius
tangent
secant

One

One.

I was the first of
them.

The numbers that is.

The numbers that is.

I was the

initiator.

Counting

and computation

started with me.

One

One

Every number

has me as a factor.

I can multiply any number

and amazingly leave
it unchanged.

One

That's me

I can divide any number

you name it

and leave it

the same.

And when you think you've

reached the end of
the numbers,

just add me to the last

and the list goes on.

I'm number one,

the first. the first.

Fractals

They call us	They called us
fractals.	monsters.
	We're young,
a rather new	
	mathematical idea.
They thought us	
	freaks,
useless discoveries.	
	Now mathematicians
know to the contrary.	know to the contrary.
Fractals are	
	the geometry of
	nature.
Snowflakes,	
ferns,	trees

clouds,
dragons

You name it

we can describe it.

We're fractals—
geometric.

We're fractals—
random.

No matter how
small
detail remains the same.

No matter how
strange
we can formulate it.

Fractals

Fractals

can paint

the universe.

the universe.

Zero

I am zero.

I am zero.

Some say I'm nothing.

I have no value.

I know to the contrary.

I'm essential

invaluable.

I'm the origin on the
number line.

The positive numbers
are to my right.

The negatives to my
left.

I'm neither negative

nor positive.

I'm zero.

I'm zero.

Centuries before I appeared

number writing was burdensome.

repetitious — confusing

I was discovered — I made the difference

in the place value system.

Now with zero there is no mix-up,

101 looks different than 11.

Without zero

there would be no

place-value system.

I am zero. — I am zero.

Add zero to any number — Multiply a number by me

the result is unchanged. — zero always results.

Divide a number by zero — Beware when dividing by me.

There is no answer. — The result is undefined.

I am zero. — I am zero.

I am nothing. — I am essential.

0

$$\text{number}+0=\text{number}$$

$$\text{number}\cdot 0=0$$

$$\frac{\text{number}}{0} \text{ is not defined}$$

$$0+0+0+0+\ldots=0$$

$$0\cdot 0\cdot 0\cdot 0\cdot 0\cdots 0=0$$

$$0\div 0=\text{undefined}$$

Fibonacci numbers

<table>
<tr><td>1</td><td></td></tr>
<tr><td></td><td>1</td></tr>
<tr><td>2</td><td></td></tr>
<tr><td></td><td>3</td></tr>
<tr><td>Fibo</td><td></td></tr>
<tr><td></td><td>nacci</td></tr>
<tr><td>numbers are we.</td><td>numbers are we.</td></tr>
<tr><td>5</td><td></td></tr>
<tr><td></td><td>8</td></tr>
<tr><td>13</td><td></td></tr>
<tr><td></td><td>21</td></tr>
<tr><td>We go on and on.</td><td>We go on and on.</td></tr>
<tr><td>Always</td><td></td></tr>
<tr><td></td><td>adding the two last of us</td></tr>
<tr><td>gives the next.</td><td>gives the next.</td></tr>
</table>

Spirals	1
pine cones	1
hexagons	2
pineapples	3
shells	5
leaves	8
flowers	13
fruits	21
phyllotaxis	34
golden ratio	55

We continually appear

in so many places
and things.

Nature

Pascal's triangle

pentagrams

infinite series

the golden

mean.

Fibo

nacci

numbers are we. numbers are we.

FIBONACCI

1,1,2,3,5,8,13,...

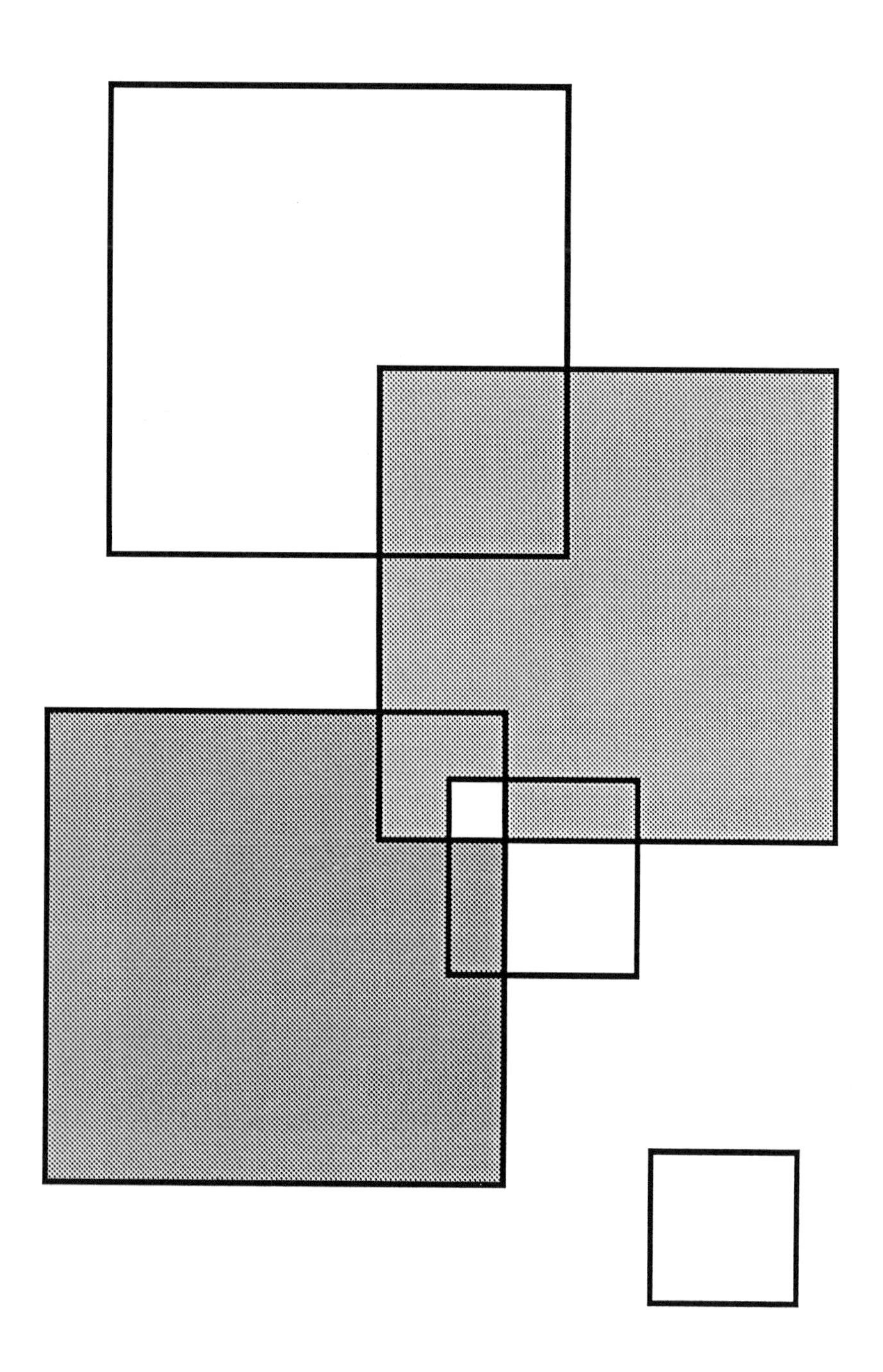

Squares

Squares	
	are not just rhombuses
We're distinct	
	not just any four sides
Four equal sides.	
Squares.	Squares.
	are not just rectangles
We're different	We're refined
Four right angles	Four equal sides
Squares—	
	we're area units
you see.	
Squares.	Squares.
	We're very special.

Operations

Operations. Operations.

Addition.

Subtraction.

Operations.

Multiplication.

Division.

We make numbers relate. We interact numbers .

We get them together. We make them operate.

and function.

We square numbers

Or take the square root.

Operations

make numbers work together—

adding

subtracting

multiplying

dividing

cubing

cube rooting

You name it we can

we can probably do it

We're

the operators.

the operators.

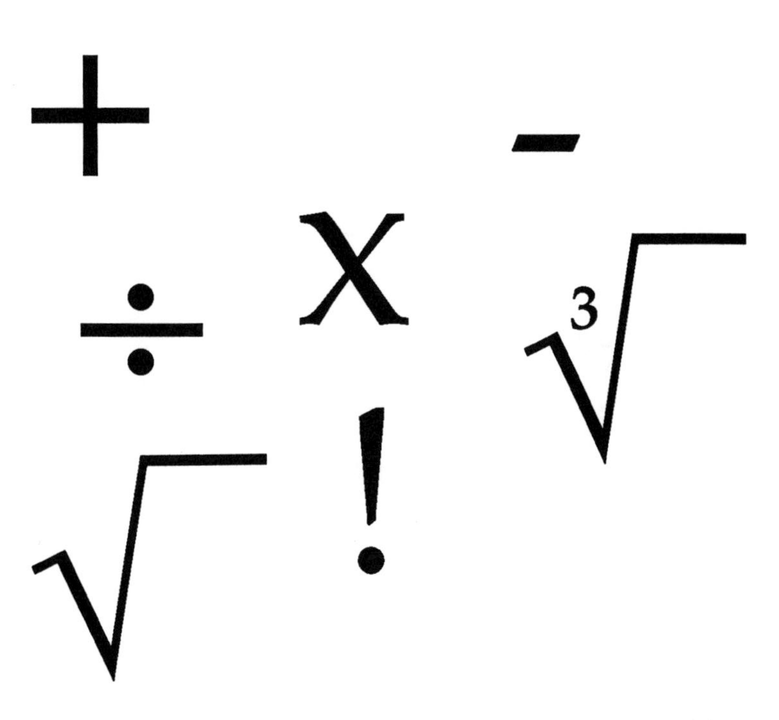

Imaginary numbers

"i"

was the first.

Confusing.

frightening
mathematicians.

What were these
monstrous numbers.

What were these
monstrous numbers.

Solutions to problems

"x" squared plus 1 equals
zero.

once thought
impossible.

We gave meaning

to such problems.

We are

the solutions

i

or negative i.

Imaginary

numbers are we.

"i" stands for

imaginary.

We appear when
a negative

We're no longer
invisible

appears
under a radical sign.

under a radical sign.

That is what did it.

That created us.

They call us

imaginary.

But we know better.

Without us

there are no solutions

to these crazy
problems.

So now there are millions

and zillions

infinitely

many imaginary
numbers

to imagine

as solutions for all
sorts of problems.

Combine us with a
real number

Combine us with a
real number

and you

end up with

complex numbers.

"a" plus "b" "i"

We're no longer confined

to one's imagination.

Imaginary in name only

We're for real!

We're for real!

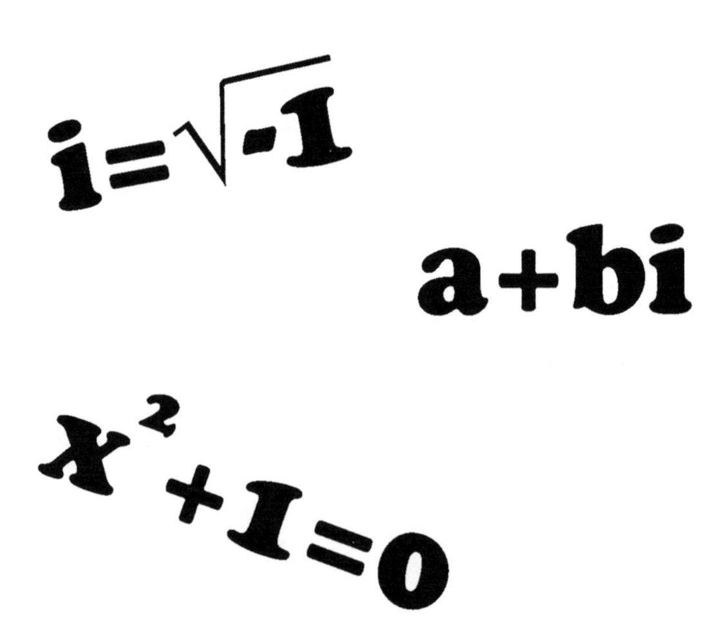

Variables

"a"

"b"

"c"

"d"

"x"

"y"

"z"

We're unknowns.

We're variables.

Representing

anything you want.

We can be found.

We can be solved.

Use us

in an equation

No matter how big
the problem

No matter how small
the problem

Let variables

help solve it.

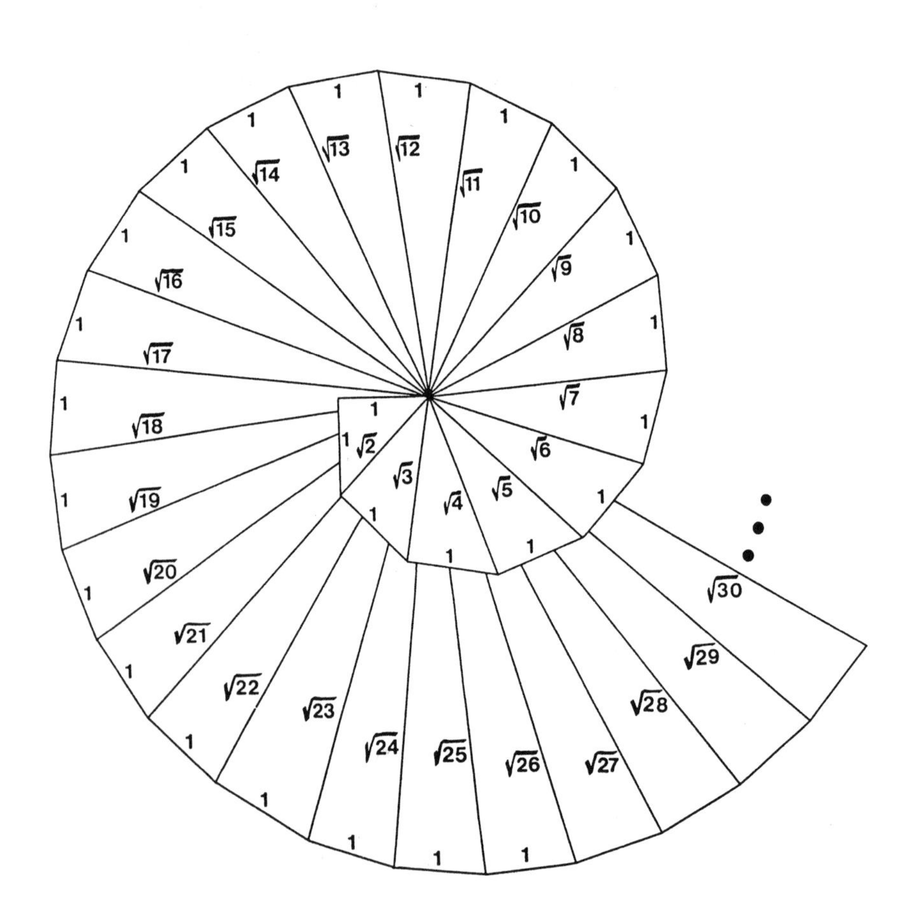

1
√2
√3
√4
√5
√6
√7
√8
√9
√10
√11
√12
√13
√14
√15
√16
√17
√18
√19
√20
√21
√22
√23
√24
√25
√26
√27
√28
√29
√30

Radicals

Square root of 2.

Square root of 3.

Square root of 4

is not one.

Square root of 5.

Square root of 6.

Square root of 7.

Square root of 8.

Square root of 9

is not.

Some call are irrationals.

We are radicals. We are radicals.

Never ending

non-repeating

decimals.

No fraction can

express any of us.

Decimals can only come

close to our values.

square root of 2	approximately 1 point four, one, four...
square root of 3	approximately 1 point seven, three, two...
But the Pythagorean theorem	
	and right triangles.
can measure	
	our lengths exactly.
Take a right triangle	
	with legs 1 and 1
The length of its hypotenuse is the square root of 2.	The length of its hypotenus is the square root of 2.
Square	
	any radical, and
the radical sign	
	disappears.
We are the strange radical numbers.	We are the strange radical numbers.
Square root of 2, of 5, of 7,...	Square root of 3, of 6, of 8,...
Never ending	
	non-repeating decimals.
Radicals are we.	Radicals are we.

The Möbius strip

The Möbius
strip

single sided.
single edged.

Once you're
on it
there's no getting
off.

Augustus
Ferdinand
Möbius
Möbius
it's
creator.

Trace a path
along its surface,

always ending

where you started—

Over

under

never

missing

any

part—

returning to

the starting point.

The Mobius strip

The Mobius strip

made

with a

half

twist.

Yet

When

cut

in half

just one

piece

remains

remains.

The Möbius strip

The Möbius strip

single sided.

single edged.

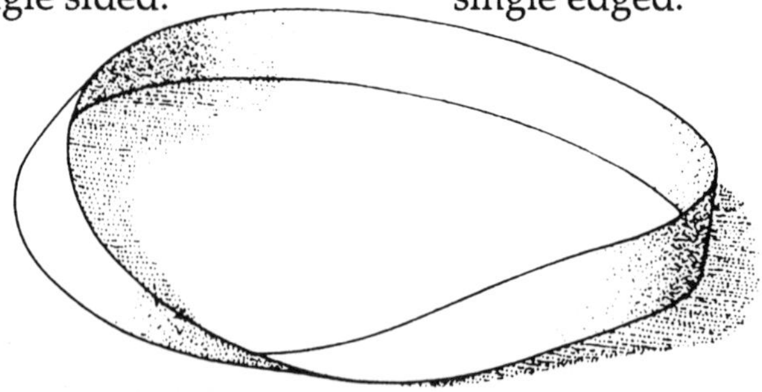

Triangles

Triangles

three-sides,

three vertices

three angles.

Triangles

three non-collinear
points.

three line segments
joined.

Triangles

scalene

isosceles

equilateral

equiangular

obtuse

acute

right.

Triangles

Triangles

large

small

always three sided.

always flat.

Two sides always

greater than the third.

Three angles always

total 180 degrees

Triangles

Triangles

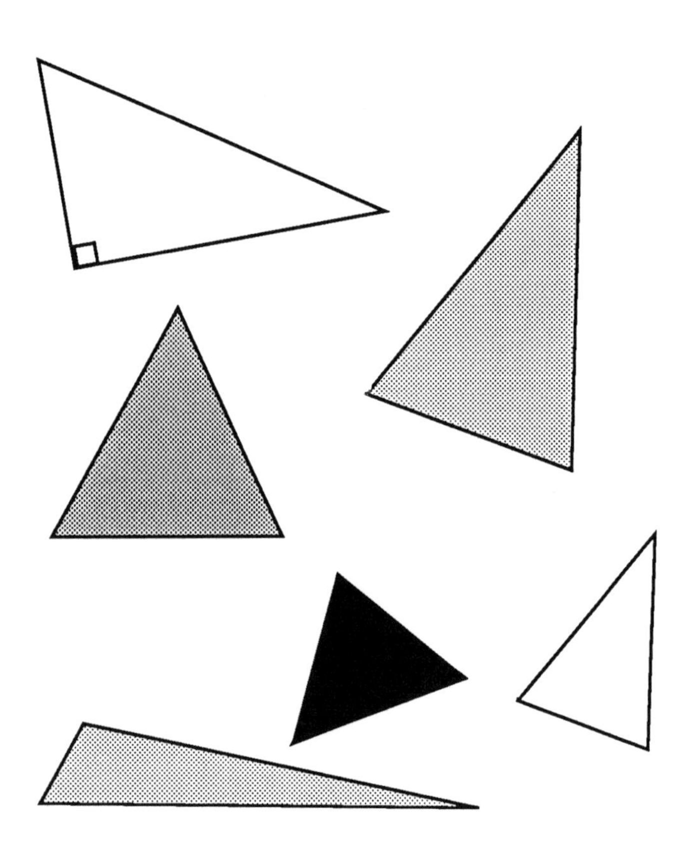

π

I was known to the ancients

to Egyptians as 3
point one, six.

Archimedes computed
me between 3
and one-seventh.

and 3 and
ten-seventy-first.

Even the Bible

referred to me as 3.

Take any circle's
circumference

and divide it by its
diameter.

What do you get?

π exactly!

π exactly!

I'm irrational

I'm a never ending

nonrepeating

decimal.

3 point

one, four

one, five

nine, two

...

Intriguing
decimals

non-repeating
decimals

tantalizing

mathematicians

for

centuries.

Hundreds of ways

have been devised

to try to come closer

and closer to me.

Polygons,

parallel lines and
needles,

infinites series

are just some of the
methods

used to reach me.

I put computers

through their paces

Now I'm out to
millions of places

Always seeking patterns

and repetitions

I have always

fascinated

and eluded

and always will!

I am π.

I am π.

π=3.1415926535
8979323846264 3
3832795028841 9
7169399375105 8
2097494459230 7
8164062862089 9
8628034825342 1
1706798214808 6

• • • • • • • • • • • • • •

Prime numbers

We like to think

of ourselves

as the basic numbers.

2

3

5

7

We can describe

any whole number
uniquely.

11

13

17

Just break down
the number

Just break down
the number

any

whole number

to its prime factorization.

No two numbers

have the same set of
primes.

We're infinte in number, yet

Only one
of us is even.

2 is the
only even prime.

The rest of us

are odd.

The only factors

of any prime

are 1

and itself.

No one can
take us apart.

No one can
factor us further.

We're not composite.

We're prime!

We're prime!

2 3 5 7 11...

The 2nd-dimension

The world of
two dimensions

Length

Width

No thickness

flat as a pancake.

Here we find

circles

squares

triangles

polygons

flat blobs

any shape

that's flat as a pancake.

Here there are no

objects with thickness.

No cubes.

No spheres.

No pyramids.

No prisms.

No you.

No me.

Here we find

only shadows

of these shapes.

Here is
pancake
world.

Here is
2-D
world.

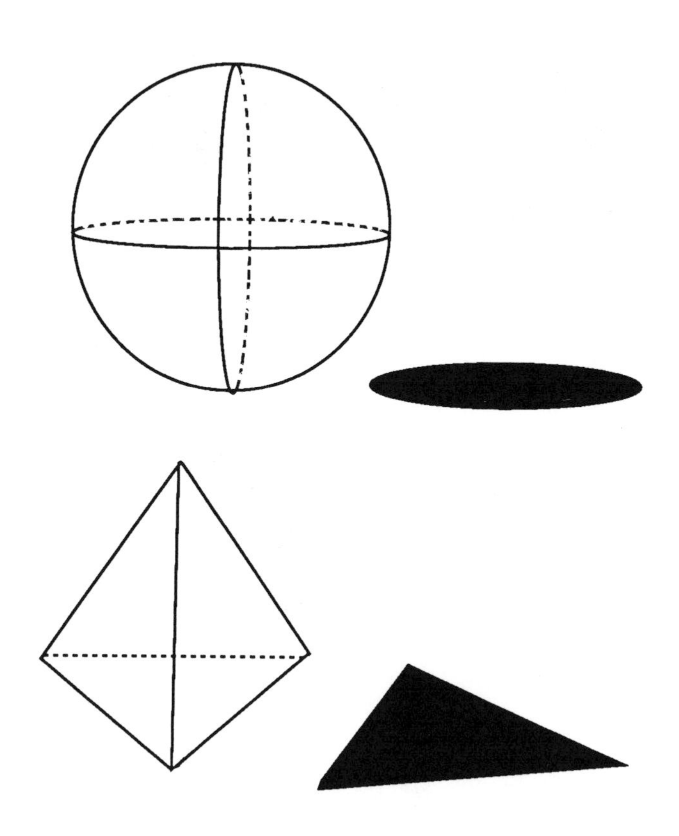

π

e

$$e^{\pi i} = -1$$

e, π and i

a poem for three vocies

I'm "e"	I'm "π"	I'm "i"
We're three		
	very unusual	
		numbers.
e and π are		
	transcendental	and irrational
i's imaginary		
But when		
	Euler	
		put us together
we came out negative	we came out an integer	we came out real
e to		
	the π	
		i
equals negative 1.	equals negative 1.	equals negative 1.

Integers

We're positive.

We're negative.

We include zero.

We include zero.

We're not fractions

nor decimals

but whole quantities.

but whole quantities.

negative 99

and positive 99

negative 2

0 and 5

We're all integers

We're all integers

Easy to list.

Easy to list.

Easy to find.

Easy to find.

We fit neatly

on a number line.

To the right of zero

are the positives
integers.

to its left the negatives.

We are integers.

Plain

and simple,

integers.
integers.

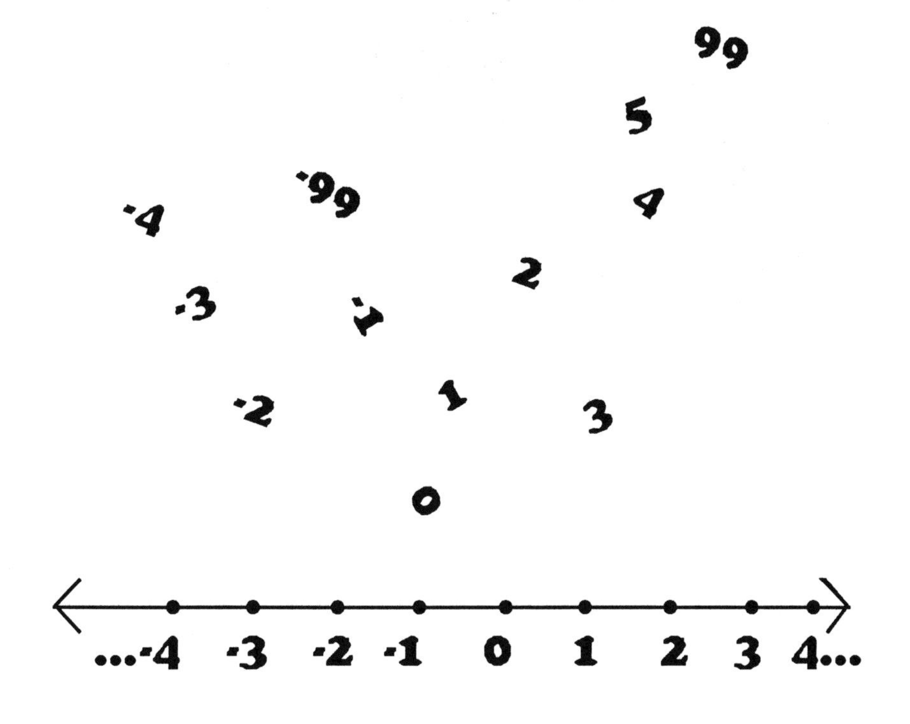

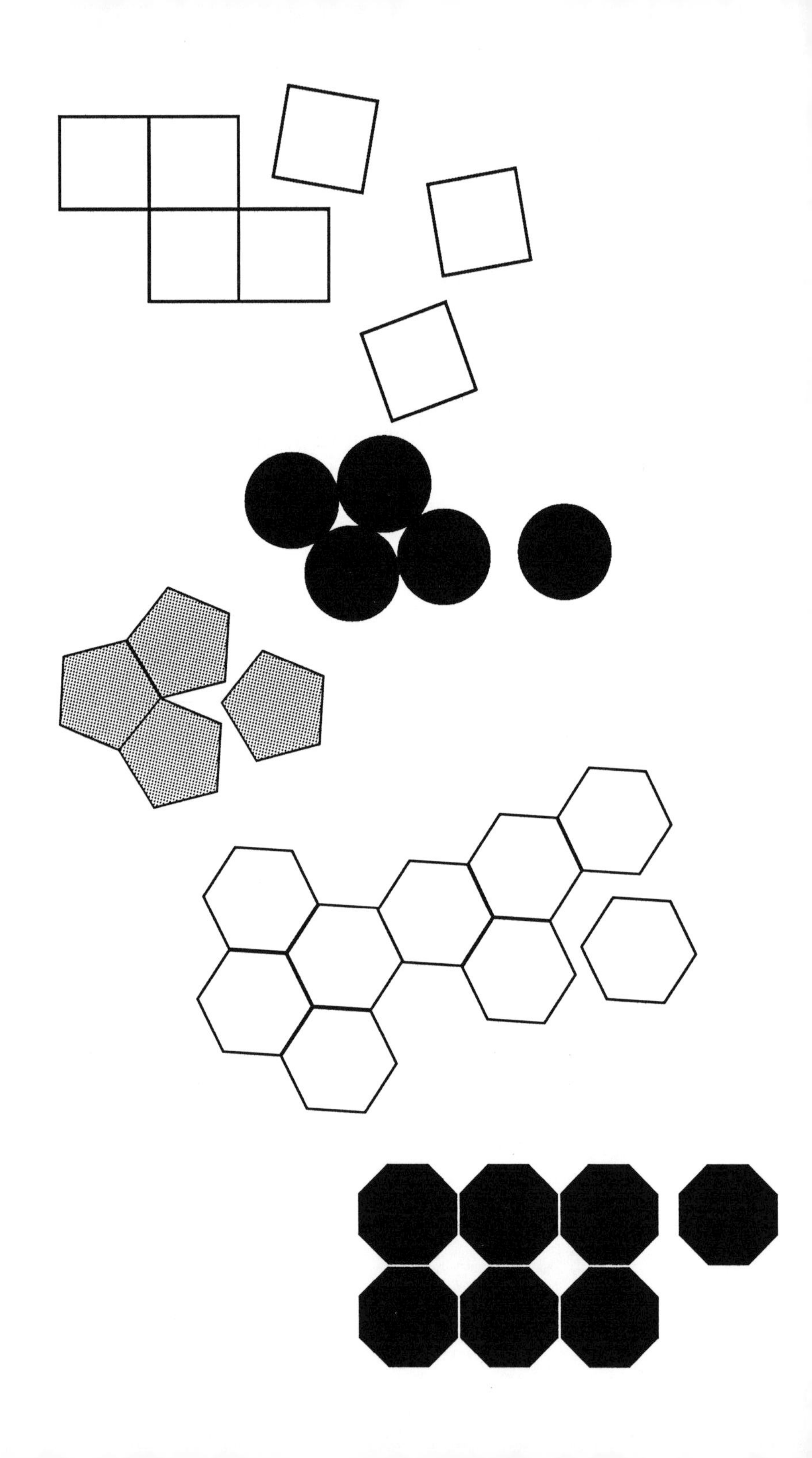

Tessellations

Tessellations
Tessellations
cover
tile

floors
walls
planes
planes

Tessellations

leave no gaps

Not all shapes

tessellate.

Squares do.

Circles don't.

What about —

hexagons
rectangles
triangles
rhombuses
yes

they do!

Not pentagons.

They leave

holes.
holes.

Tessellations
Tessellations
leave no gaps.
leave no gaps.

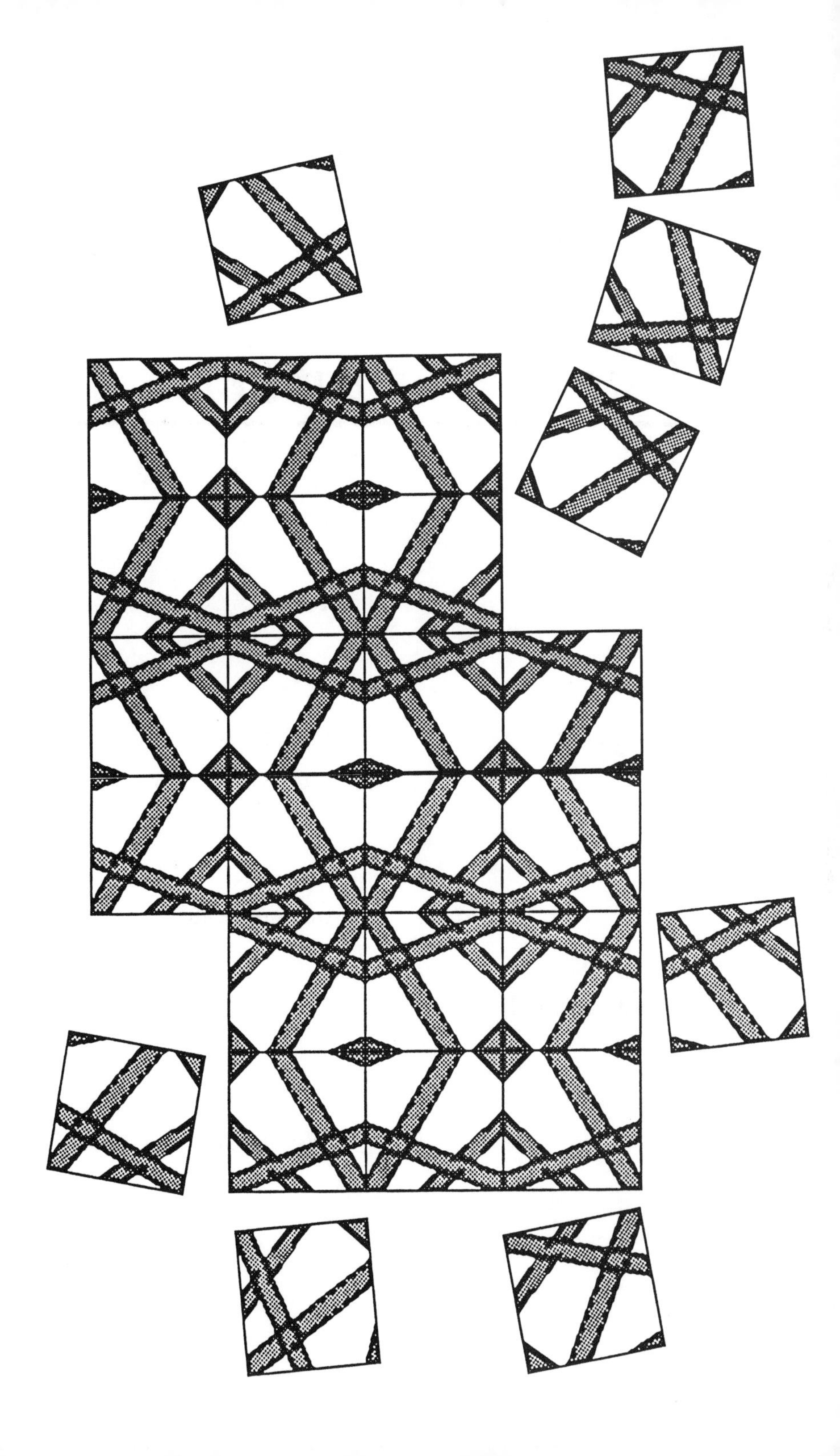

The even numbers

2

4

6

8

We are even

We are even
and were great.

2 is

our generator.

Common to all.

Without 2

We're odd.

2, 4, 6, 8,
10, 12, 14,
16, 18, 20,
22, 24, 26,
28, 30, 32,
34, 36, 38,
40, 42, 44,
46, 48, 50,
52, 54, ...

Proper fractions

We are proper

fractions.

Ever so correct.

Never out of line.

Never larger on top

than on bottom.

Never hanging

around with whole numbers.

Our numerators

and denominators

are simply whole numbers.

are simply whole numbers.

Numerators always
smaller than
denominators.

We are proper
fractions

Denominators always
larger than
numerators.

We are proper
fractions
that is.

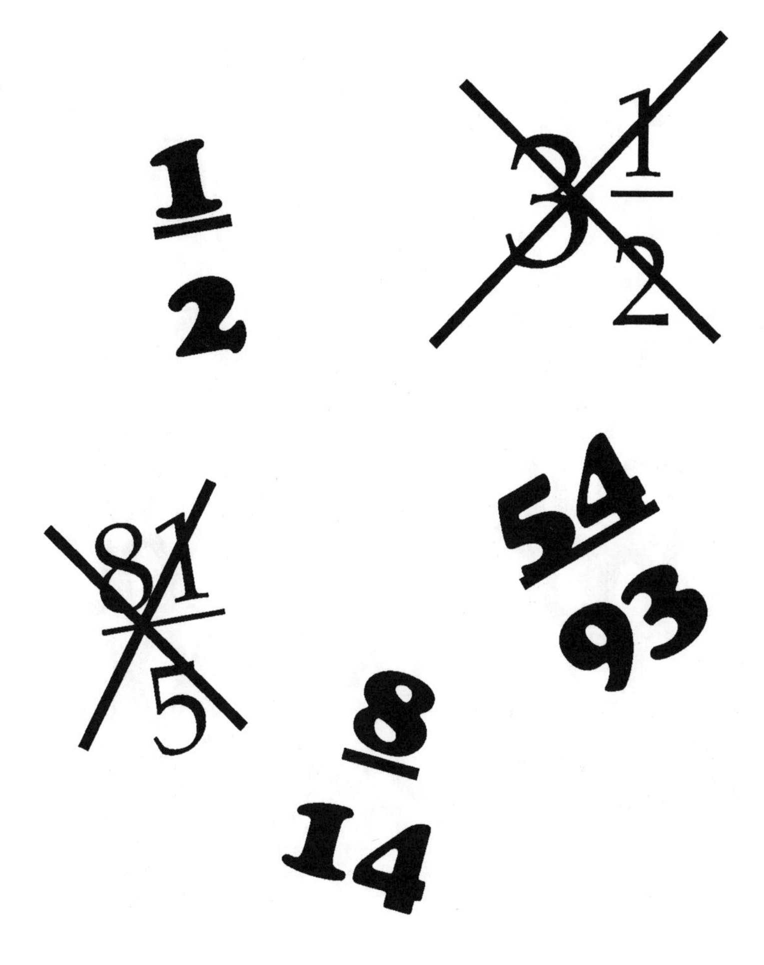

Googols

Googols	
	Googols
So big.	So large.
A number named	
	by one so young.
No one can	
count to a	
googol.	googol.
	Few people can
	use a
googol.	googol.
Takes so long	Takes such space
to write	to write

10^{100}

a googol's shorthand

10^{googol}

is a googolplex

a googol.	a googol.
One followed	
	by one hundred
zeros.	zeros.
Googols—	Googols—
Powers of ten	Powers of ten
cut it to size.	cut it to size.
Ten to a hundred	A googol's shorthand.
Googols.	
	Googols.
They're	They're
so big,	so large,
so great.	so great.
But	
so small	so small
next	
	to
googol	
	plexes.

Infinity

Infinity
No simple idea.

Infinity
No small idea.

Ever expanding.

Ever contracting.

Where does it end?

No where.

Some only think

of it as enormous.

It is endless

never-ending.

While without end,

infinity can be
confined.

Infinity.

Infinity.

Confusing.	Exciting.
Mind-boggling	Stimulating
concept.	idea.
Cantor categorized	
the transfinites	the infinities
aleph-null	
	and all the rest.
Find infinity	in sets of numbers
the naturals	the integers
the whole	the even
the primes	the odd
the points on a line	the points on a circle
the points	
	between any two
	points
the number of numbers	
between 1	everything
and 2.	throughout
Infinity!	the universe.

The golden mean

I'm a very special	Quite extraordinary
ratio	ratio
Not just any old	
	fraction but
the golden mean	the golden mean.
Discovered by	Not discarded
the ancients.	by moderns.
Use me to form	Look for me in
a golden rectangle	regular pentagons
golden triangles	points of pentagrams
equiangular spirals	shells, pine cones, natural
	shapes
architecture	sculpture and art
Parthenon	Phidias

da Vinci
Dürer
Dali
created

Mondrian
Seurat
Bellows

dynamic symmetry

by using golden means.

So special a ratio
is "phi"—
the golden mean.

My value is given
the name—
the golden mean.

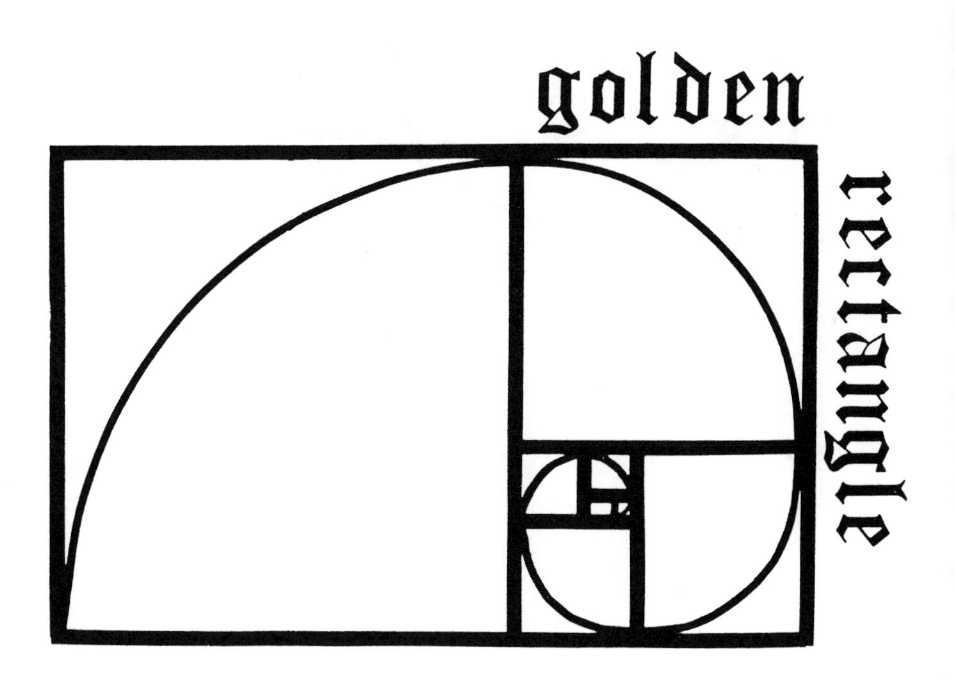

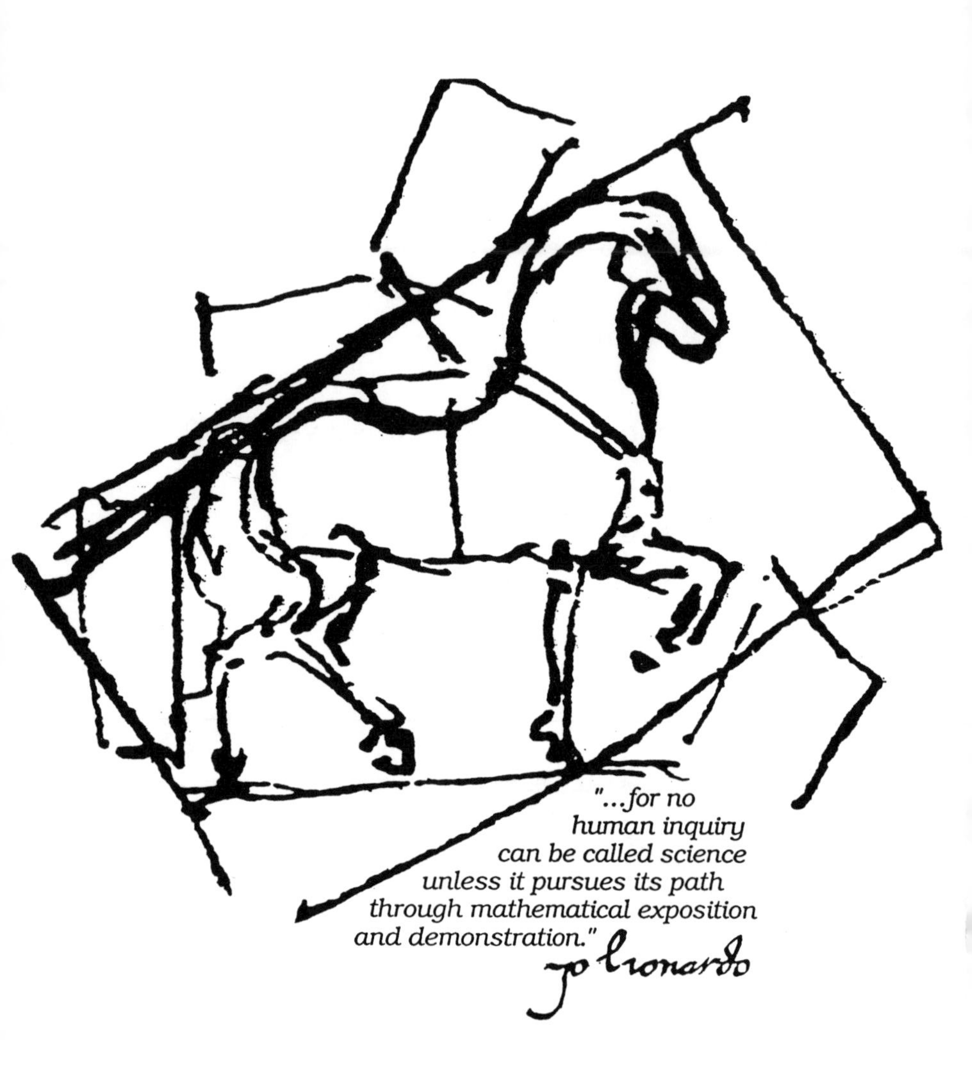
"...for no
human inquiry
can be called science
unless it pursues its path
through mathematical exposition
and demonstration."
Io Leonardo

About the author

Mathematics teacher and consultant Theoni Pappas received her B.A. from the University of California at Berkeley in 1966 and her M.A. from Stanford University in 1967. She is committed to demystifying mathematics and to helping eliminate the elitism and fear that often are associated with it.

In addition to *Math Talk* her other innovative creations include *The Math-T-Shirt, The Mathematics Calendar, The Children's Mathematics Calendar, The Mathematics Engagement Calendar,* and *What Do You See?*—an optical illusion slide show with text.

Pappas is also the author of the following books:

Mathematics Appreciation
The Joy of Mathematics
More Joy of Mathematics
Greek Cooking for Everyone
Math Talk
Fractals, Googols and Other Mathematical Tales
The Magic of Mathematics
The Music of Reason
Mathematical Scandals
The Adventures of Penrose—The Mathematical Cat
Math for Kids & Other People Too!

Her most recent titles include *Math-A Day* and *Mathematical Footprints.*